Vedic Mathematics

Vedic Master-formula for Powers Roots

(A Revolutionary Concept That Changes the Way We Learn and Teach Mathematics)

Jack Alston

Published By **Phil Dawson**

Jack Alston

Vedic Mathematics: Vedic Master-formula for Powers Roots (A Revolutionary Concept That Changes the Way We Learn and Teach Mathematics)

ISBN 978-1-998038-36-7

Legal & Disclaimer

The information contained in this book is not designed to replace or take the place of any form of medicine or professional medical advice. The information in this book has been provided for educational & entertainment purposes only.

The information contained in this book has been compiled from sources deemed reliable, and it is accurate to the best of the Author's knowledge; however, the Author cannot guarantee its accuracy and validity and cannot be held liable for any errors or omissions. Changes are periodically made to this book. You must consult your doctor or get professional medical advice before using any of the suggested remedies, techniques, or information in this book.

Table Of Contents

Chapter 1: Vertically And Cross-Wise]

Let's review the general method that we use to do mental math. If you have taken algebra, you already are acquainted with the form that is used. Multiply your two "factors" as though they were in algebraic form. In Vedic math this is called "Vertically and Cross-wise". It is in fact one of the 16 Sutras of Vedic Math. This method allows you work from left to right. First, do the vertical multiplication of the first two digits, then cross multiply and add, finally, do the vertical multiplication of the last two digits.

If you were multiplying 23x12, you would start out by multiplying 2*1, then (2*2)+(3*1), finally, multiply 3x2. The components to your multiplication would be 02/07/06. Your answer would be 276. Using this method you can multiply any algebraic expression. What is particularly exciting about this method is that you can write your answer down from left to right on a single line.

This fact alone gives you a huge advantage when multiplying compared to the grade school method you learned. You can assemble the parts in your head as you go. The method also allows you the opportunity to look at the relationship of the parts to each other. The nature of the relationships is a huge part of mental math. Those relationships allow you to do shortcuts and give you a math agility most people confined to the grade school method never develop. Indeed, I almost never use the grade school method you learned to do math. I do use a pencil and paper, but always think in terms of components and relationships. Very simply, you can't beat a system that involves the use of your mind as opposed to the mindless use of a single method such as the grade school method.

With two and three digit multiplication you will always have three components to assemble. If you put them together in order from left to right, you are basically writing the answer down as you calculate. In the case of three digit multiplication you split the three digits into a two digit number and a one digit number. Generally, you would split both numbers the same way. For example, if you want to multiply 112x103, you could split the 112 into 01 and 12. Then split the 103 into 01 and 03. Now proceed to do your vertical and crosswise calculations. The components would become 01/15/36. Your answer would be 11536.

Conversely you could have split 112 into 11 and 02. You could have split 103 into 10/03. In this instance your components would be 110/53/06= 11536. Three digits is no more difficult than two digits, you have the opportunity to split the number in a fashion that is easiest for you. The only thing that is different is that you will have two one digit by two digit multiplications to do and add together to get your cross product.

In that sometimes that cross product calculation is not easy, we have a myriad of ways to make it simpler. Finding the cross product is the hardest part of mental math. For one thing when the number one is involved we have another Sutra we can apply. That one is called By One More Than The One Before. This book is largely about the application of that sutra. The method works to multiply numbers close to a base of 1, 10, 100, or more, you note the incremental amount it is over or under a base of one, ten, or more. Then you work off of that. In the case of the problem we just solved we could say one number was 12 units greater than 100, and the other was 3 units greater than 100. Now we either add 3 to 112, or add 12 to 103. Either way we get 115. Now we just tack on the product of 12*3=36. Our answer is 11536.

We will be going into this in much more detail in the ensuing chapters, it is quite simple, and effective for situations where you are close to a base especially. View it as a different tool that will come in handy when you want to do a quick calculation.

Being close to a base was one of the topics I didn't get to in my first book Using Fractions to Multiply. One of the reasons I didn't cover that topic was that I had already thrown so much out there in regards to math. I wanted to confine that book largely to a single topic. That topic was fractions, ratios and the use of those to help you become a better mental math calculator.

Along the way we covered far more than that. This was largely because you can't talk about ratios and fractions without touching upon a whole bunch of other things. In this book we are going to see how the identity property is way underrated as to it's power and application to mental math. The identity property of multiplication means that you can multiply any real number by the number one and not change its value.

You would say, and I would have to concede that this definition seems lame and worthless. By the time we get done working on multiplication close to a base of one, ten, or one hundred you will appreciate how useful this simple property is.

Along the way we are going to touch on a lot of other things again. Hopefully, I can touch upon some of the same items we talked about in the first book in a way that briefly covers those items , and yet elaborates on them as well. Just as we did in the other book we will start off by discussing single digit by double digit multiplication, this time from the perspective of the identity property.

Then we will go from there to see how it is bidirectional and symmetrical. We will see how the technique we're going to learn can be combined with the use of ratios. For sure we will be elaborating even more on the use of digit sums to multiply, and synchronized ratios. We will also look at a couple of my favorite numbers nine and 11, and we will elaborate on their properties even further.

One of the key skills you need for mental math is the ability to multiply a double digit number by a single digit. There are all kinds of shortcuts that you can use to help you do this in your head. Multiplying a double digit by a single digit involves linking two multiplications together, one offset from the other by a multiple of 10.

For example, if you wanted to multiply 9x31. You could easily do that in your head by multiplying 9x3= 27. Now you link that 27 with the product of your second calculation 9x1, which equals nine. The numbers you are linking then are 27/09. Clearly, you can see that the zero was important to help you place keep. Our answer is 279. Note the identity property came into play on this one, in the form of 9x1= 09.

Since ratios and proportions are so important to mental math, my suggested way to approach this problem is not to just grind out the numbers. Just take a second and think of 31 as a fraction. That fraction would be 3/1 or the whole number 3. If you look at it from the other direction. The fraction is one third. What I'm getting at is that if you know one of the two multiplications and the relationship of the numbers to each other. You don't have to do the second multiplication.

In this case, the lead numbers 27, in the second number is 27 divided by three, or nine. Tack the nine onto the 27, and you have your answer 279. This is a subtle departure from what you were doing previously. But the distinction is very important. What you're really doing is recognizing a pattern or relationship first, and then working off of that. Considering your double digit number as a fraction is the starting point for using fractions to multiply.

When the relationships are real simple this works pretty easily. Anything that begins or ends with one makes a whole number and it's reciprocal is a easy fraction. The math is about as simple as you can get. Later on we will find out that double digit multiplication that begins or ends in one is just as easy. Relationships where one number is twice the other, or three times, are also quite simple.

For example 48x9. Nine times four equals 36. The second number is merely two times that amount. So you're linking 36/72 to get 432.

Nine as a multiplier is particularly interesting. Perhaps you've heard that casting out nines is a good way to check your answer in multiplication. It's actually even more useful than that. Anything you multiply by nine will have a digit sum of nine. The digits in 432 add up to nine. The components add up to nine too. Nine combined with the number 11 also has some interesting characteristics.

If you are in middle school you have taken, or are currently taking algebra, or will take algebra. When you run into algebraic equations where (x +1) is a factor, it will always turn out that the sum of the odd coefficients will equal the middle coefficient. The mental math equivalent of this would be multiplying a number by 11. Anything you multiply by 11 will have middle coefficients, or components that add up to the sum of the odd coefficients.

11X23, for example is 253. The two and the three added together equals that middle number five. Sometimes when the numbers are larger, this fact is hidden because we go over 10 in the middle and carry one of the digits over to the lead number. It is also true that (x-1), which is the algebraic equivalent of nine has the characteristic of always giving you an answer that adds up to nine.

In algebra, this manifests itself in a distinct way. The odd coefficients added together with the middle coefficient always equals zero. The algebraic equivalent of multiplying 48*9, would be multiplying the factors of the two equations (4x+8) * (x -1). The expansion of this problem, would give you (4x2+4x-8) notice how the sum of all the coefficients equals zero?

The other easy tip I wanted to mention in regards to nine is about multiplying a two digit number by it that has a digit sum of eleven.

For example 9*92, or 9*83. The numbers that you are linking together are reciprocals of each other. There is no need to do the math so to speak beyond the first multiplication. In the first case you are linking 81/18, in the second case you are linking 72/27. The two outside "wings" are equal, the inside "body" is twice the last digit. So you have an answers of 828, and 747.

This is not unlike multiplying two numbers together that are reciprocals. 72*27 would have components of 14/53/14. Again, the outside wings are identical, they are the product of 7x2. The middle body part is always the sum of the squares of 2 and 7. In algebra whenever you see first and last coefficients that are identical in value and sign you should look for a possible reciprocal relationship.

From an algebraic standpoint 27x72 is the same as $(2x+7)*(7x+2)$, where x=10. The expansion would be $14x2+53x+14$. If you were factoring this expression you could rule out a lot of things very quickly. The factors of 14 and 1 are not worth considering. The factors of 7 and 2 are the ones to consider 7-2=5, 5-2=3. There you have it by observation.

From the factors you also get the roots. The whole point being you want to contemplate a relationship, and not get bogged down grinding out the factors. From a fraction standpoint you are looking at 7/2+2/7 which equals 49+4/14. Everything you need to get your answer is contained in that expression. The components 14/53/14= 1944.

There is a whole class of problems, namely two digit squares where fractions are very important. For example, if you wanted to multiply 12x12. You could do it by considering each number as a fraction. In effect, you would be adding 1/2+1/2 to get a whole number of one. That means in this case that there is a one to one relationship of the product of the denominators to the middle component of our calculation. We know the 2x2 equals four, just knowing that and that one times one equals one leads us to know that 12x12 involves linking together the following three components 01/04/04 =144. Normally to get that second component, or middle coefficient you would have to cross multiplying and add (1*2)+(1*2)= 04. You can skip all that laborious work by just knowing that you have a one-to-one relationship with the product, 2x2. If the problem were turned around in you are looking to find the square of 21. You could again look 21 as a fraction. 2/1+2/1= 4.

This time the middle coefficient is exactly equal to the product of 2x2, the lead number product. We did one multiplication and got to the three components we need to do the math in our head. In this case we will end up linking together 04/04/01= 441.

Vedic math is partially an algebra-based math that allows you to mentally do significant calculations in your head. Many of the rules for manipulating the numbers are based on little rhymes are sutras. There are 16 of them altogether., and there are some sub sutras as well. I urge you to google vedic math, visit some of the web sites out there on the subject. It is fascinating how easy this method makes math.

You are not limited to fractions that add up to one to use fractions to multiply. The can add up to other whole numbers. 42X42. For example, would give you the whole 8/2+8/2= 4. Or looking at the fraction the other way would give you ¼. So immediately you know that you are linking 16/16/04 to mentally get 1764.

The use of fractions can be extended to include numbers that are not squares. For example, 13x23. In fraction form you are adding 1/3+2/3 equals one. That means your components are 02/09/09= 299. The use of fraction form does not limit you only to the whole number one. You could have other whole numbers, you can have whole numbers plus a fraction and still quickly come up with components. It really pays to look at the numbers as a fraction before you dive in and start trying to arrive at the answer.

We can take the 13x23 example and extend the use of fractions to include multiples. For example 26 is still the fraction 1/3. 26x23 would still give you two fractions that add up to the whole number one. In this case you don't really have to add the fractions, just recognize that they add to one so now you know that your middle component is still the same as the product of the denominators 6x3. The components and answer are 4/18/18= 598.

In the case of three digit multiplication you just look at the three digit number as an improper fraction, or a fraction with a two digit denominator. For example, if you wanted to multiply 412x812. You could look at the sum of the fractions

(8/12)+(4/12). Since they add up to one you know without doing any math that the middle component is going to be the same as the product of the denominators multiplied together. So now you just have to put together 32/144/144= 334,544. The number 612 you could square in your head to get 374544. you just did real three digit multiplication in your head, and it was real easy!

So there you have it a quick review. Nine and eleven pass genetic traits onto, or into everything they multiply with. Nine has a secret reciprocal relationship with all the two digit numbers that have a digit sum of 11 that it multiplies with. I might add that nine has a lot of secret relationships beyond that. You can predict the second product of any two digit number you multiply by nine by just looking at the digit sums. Reciprocals have a special pattern as well. Beyond that there are the older shortcuts everyone knows about.

You can square any two digit number that ends in five by multiplying the first digit by one more than it's value, and tacking 25 onto it. 25X25 would be 3*2, with 25 tacked onto it to get 625. 35x35 would be 4x3 with 25 tacked onto it to get 1225. This same "by one more" sutra can be used when you multiply any two digit numbers together that have the same tens base and units that add to 10. For example, 81x89 would be 9x8 plus tack on 09 to get 7209.

The same units and base approach can be extended to include some very powerful methods. Notice how "by one more" comes into play on this example. If you wanted to multiply 21x39 you could note that the units digits add up to ten and the larger tens digit is one more unit greater than the small digit. In this case, your cross product needs no calculation it is in fact right under your nose. It is in fact the smaller number itself.

Anytime you have units that add to ten and tens units that are one unit apart the smaller number itself is the cross product. In the 39x21 example skip doing any fractions or cross multiplication to get the cross product. In this case the components are 6/21/09= 819. If you were multiplying 82x98 you can virtually write your answer down from left to right as 72/82/16= 8036.

Notice that we can also look at 39x21 and borrow a set of ten from the larger number. By doing that we make the 39 into 2/19. Now the two numbers have the same base of 2 but the units digits add up to 20 now:

2 19

2 01

08/ 19= 819

In this case, we multiplied our base by 4 more and tacked on the 19.

If the question had been to multiply 49x21 we could do a similar procedure and turn 49 into 2/29. Now we have 2/29 and 2/01. The base is the same and the units add up to 30. We multiply the base by the base of 20+30=50. So our new lead number is 10 with 29 tacked onto it. We have expanded this little shortcut to be very useful even when the initial precondition of having the same base is not immediately apparent.

In the three digit world the same principle works also. If you wanted to multiply 248x252 you could note that 48 and 52 add to 100. By one more would mean you multiply the 2 by 3 to get 6. Now tack 48x52 onto that. Multiplying this in your head is pretty easy, the two numbers add to 100 and average out to be 50. Each number is two units from 50. 50x50=2500. Now subtract adjust the 2500 by subtract 2x2 from 2500 to get 2496. Tack 2496 onto 6 to get 62,496 as your answer.

If the two numbers being multiplied were 219x211 we could also multiply 21x22 to get 462 and tack 09 onto that to get 46,209 as our answer. Really all that happened by going to three digits is that we expanded the opportunities we have to apply our shortcuts to finding a solution. It's the old story of difficulty creating opportunity.

The bi-directionality principal applies too. We will be reading about that in the next chapter. Basically it means that what works in one direction works in the other. If you were asked to multiply 39x21 backwards you would be multiplying 93x12. You could note that the tens units add up to ten and the units are one unit apart. So now your calculation of components would be 09/21/06= 1116.

My cross product is still 21, I just had to read it backwards or from right to left. It's still there though. This one is so powerful it is incredible. We doubled it's power with the fact of bi-directionality, and we can extend it to three digit numbers also. Consider for a moment what happens when the total of the units column is not ten. Very easy to adjust for. Each unit over ten or under ten is 1/10th. We will definitely be looking at this in a chapter all by itself. So without further rambling let's look at more math tools and opportunities.

Chapter 2: By One More

In actuality, you never really needed to learn the multiplication tables past 5x10. Let's say you wanted to multiply 9x8 but didn't know your tables that far. You could do the multiplication anyway, by taking the compliment of these two numbers. That would be one and two, respectively. Since these two numbers are compliments, they represent the deficit from 10 that we have. The word deficit implies that they're both negative numbers in regards to the base 10.

If you want you could represent these two numbers on your fingers. Put one and first finger of your left hand, and two on the first two fingers of your right hand. In any event, we can now subtract two from nine, or one from eight. Either way, we get seven. Hold that seven in your mind for a second, while you multiply the two numbers on your fingers together. Those two numbers are -1 and -2. Multiplied together they equal +2. Let's tack that two onto the seven to get 72. 72 is of course the product of multiplying nine times eight. What just happened? We actually multiplied 7x8 without using the times tables beyond multiplying one times two.

Let's do the same thing with 7x6. Put -3 on your left fingers. Put -4 on your right fingers. Now subtract -4 from seven, or -3 from six. Either way, we get three. Multiply the numbers on your fingers together to get 12. Now put 3/12 together as though three were 30, and we were adding 12 to it. And give us 42. 42 is in fact 7x6.

All this is well and good, but you already know times tables past five. So what value is it? Well, this is the starting point for learning to use the identity property to help you multiply. In the background there is the fact that we were working with the number 10 as a base. The numbers 1, 10 , 100, 1000, and indeed 10,000 and beyond are all subject to the identity property. Let's leverage this concept a little more.

If you wanted to multiply 6x14, you could do it in a similar fashion to what we just did. Note that six is four under the base of 10. 14 is four over the base 10. We have a common base, let's put four on the fingers of one hand, and four on the fingers of the other hand. Now we can either add four to six, or subtract four from 14. Either way you get 10. Now multiply those numbers on your fingers together. 4X4= 16. since one number is over a base in the other is under a base. We are dealing with a negative number, times a positive number. That resultant 16 is a negative number in this case. Subtract 16 from 100 and you have 84 , which is in fact 6x14.

Let's do another example. This time, let's try 8x17. One number is two under a base of 10. The other is seven over a base of 10. Put the number two on your left fingers. Put number seven on your right fingers by using your thumb as a five plus 2 more fingers. Now either subtract two from 17, or add 7+8. Either way you get 15. Hold that 15 in your head for a second and multiply the two numbers on your fingers together 2*7= 14. So now we have 15/-14 to deal with. That would be the equivalent of 150-14= 136.

Now let's try an example where both numbers are over the base of 10. Let's multiply 13x19. Put three on the fingers of your left hand, and nine on the fingers of the right hand. We are using a Chinese Abacus method with our fingers for numbers over five. In this case your thumb is the five bead and all your remaining fingers extended represent nine. Since everything is over a base of 10 everything is additive, or positive. We are adding 9 to

13, or adding 3 to 19. Either way , we get 22. Now we link the 22 with the product of our fingers. That would be 3*9=27. So we are linking 22/27 to get 247. All this can be done mentally with a little help from our fingers. Is this pretty cool or what?

Let's do one more where both numbers are over a base of 10 just to make sure we have this method down. Let's do (19*19). This should be pretty easy. Both numbers are nine over a base of 10. Add 9 to 19 to get 28. That is the first half of the answer. Now we link it with the product of the nines we have on the fingers of each hand. 9*9 equals 81. 81, then is the second half of our answer. We have to link 28/81= 361. We just squared 19 mentally.

I don't know if the Karen Carpenter song We've Only Just Begun is still a standard at weddings these days, but as the song says we have only just begun. Soon you and math will be together in bliss! Firstly, let's do some exercises using the techniques we just learned. You can check your answers with a calculator if you like. I would prefer that you use fractions, and some of the other methods we learned in my previous book to check your work. Below is a example template for that.

Remember that when you look at the fraction you are looking at the relationship of the cross product to either the first vertical product, or the second vertical product.

Exercise 1:

Base Left Right Result Answer

1.   6x6 2 -4 -4 16 2/16

2.   6x7 3 -4 -3 12 3/12

3.   6x8 4 -4 -2 08 4/08

4.   6x9 5 -4 -1 04 5/04

5.   7x7 4 -3 -3 09 4/09

6.   7x8 5 -3 -2 06 5/06

7.   7x9 6 -3 -1 03 6/03

8.   8x8 6 -2 -2 04 6/04

9.   8x9 7 -2 -1 02 7/02

10.   9x9 1 -1 -1 01 8/01

11.   6x12 8 -4 2 -8 8/-08

12.   6x13 9 -4 3 -12 9/-12.

13.   6x14 10 -4 4 -16 10/-16

14.   6x15 11 -4 5 -20 11/-20

15.   7x16 13 -3 6 -18 13/-18

16.   7x17 14 -3 7 -21 14/-21

17.   7x18 15 -3 8 -24 15/-24

18.   7x19 16 -3 9 -27 16/-27

19.   8x12 10 -2 2 -4 10/-04

20.   8x13 11 -2 3 -6 11/-06

21. 8x14 12 -2 4 -8 12/-08

22. 8x15 13 -2 5 -10 13/-10

23. 9x16 15 -1 6 -6 15/-06

24. 9x17 16 -1 7 -7 16/-07

25. 9x18 17 -1 8 -8 17/-08

26. 9x19 18 -1 9 -9 18/-09

27. 13*16 19 3 6 18 19/18

28. 13*17 20 3 7 21 19/18

29. 13*18 21 3 8 24 21/24

30. 13*19 22 3 9 27 22/27

31. 14*16 20 4 6 24 20/24

32. 14*17 21 4 7 28 21/28

33. 14*18 22 4 8 32 22/32

34. 14*19 23 4 9 36 23/36

35. 15*16 21 5 6 30 21/30

36.   15*17 22 5 7 35 22/35

37.   15*18 23 5 8 40 23/40

38.   15*19 24 5 9 45 24/45

39.   16*16 22 6 6 36 22/36

40.   16*17 23 6 7 42 23/42

41.   16*18 24 6 8 48 24/48

42.   16*19 25 6 9 54 25/54

43.   17*17 24 7 7 49 24/49

44.   17*18 25 7 8 56 25/56

45.   17*19 26 7 9 63 26/63

46.   18*18 26 8 8 64 26/64

47.   18*19 27 8 9 72 27/72

Chapter 3: Bi-Directionality

One of the great things about mental math is that the calculations can generally be done from either direction, and a certain symmetry generally is present. What is so great about this is that the methods, shortcuts and techniques you have at your disposal then are twice as useful or powerful.

Let's consider for a moment some of the exercises we did in the previous chapter. One of them, for example, was 16x16. In this one. Each number was six units over a base, so we headed 6 to 16 to get 22, and linked that to 6*6= 36. The result of the calculation was 22/36 = 256. If we were to turn that number around, we would be calculating 61x61. Could we use the same method?

The answer is yes! Now we are saying that each number is 60 units over a base of one. From there we had 60 and 60. We could even use our fingers on this one too, by letting each finger represent 10 units, and your thumb representing 50. Now we just add the two sixes together to get 12, and multiply that answer by 10 to get 120. Then we simply tack 36/12/01 together to get 3721.

The major point about this being that anything ending in one is fair game for the same technique we used for something that started with one. It really is no more difficult to multiply 16*16. than it is to multiply 61*61.

Let's try another calculation based upon the same thought process. You want to multiply 91x61. You could do it by putting 90 on the fingers of one hand and 60 on the fingers of the other hand. First multiply them to get one link, and then add them to get the second link. The last link is simply 01.The links together would be 54/15/01= 5551. That calculation was about as simple was any calculation you ever do. It's a nice ,simple, straight left to right calculation.

We can even apply this to three digit numbers that end in one. For example, if you wanted to multiply 121 times 141 you could easily do it using this method is long as you knew that 12x14 was 168. The components would be 168/26/01. The answer would be 17,061.

You could even do a two digit by three digit multiplication this way. If you wanted to multiply 81x121 for example. As long as you knew that 8x12 was 96 this should be very easy. Your components would be 96/20/01. Your answer would be 9801. Now let's do some exercises that use the same principle.

Exercise 2

 base sum links answer

48. 11x21 02 03 02/03/01 231

49. 11x31 03 04 03/04/01 0341

50. 21x41 08 06 08/06/01 0861

51. 31x51 15 08 15/08/01 1581

52. 41x61 24 10 24/10/01 2501

53. 51x61 30 11 30/11/01 3211

54. 51x71 35 12 35/12/01 3621

55. 61x71 42 13 42/13/01 4331

56. 71x81 56 15 56/15/01 5751

In other cases it may be better to look at incremental amounts over or under a base. Personally anything with a common denominator of one whether read from left to right, or right to left is very hard to resist as a fraction. As long as you realize that 01*01= 01 and skip actually actually calculating it out.

If you think about it this is similar to adding fractions that have a common denominator of one. We did the same calculation in those cases by multiplying the leading digits, adding the fraction to get a whole number, and then squaring 1*1 to get the last digit of the link. Only now the story is different, we are talking about being over a base of one as opposed to having a fraction that has a denominator of one and is therefore a whole number.

Essentially we are describing the same operation with two different stories. Which one you prefer is up to you, the big deal is to get the math right. The same applies in the other direction. If you look at 16*16 as two fractions you can work from left to right and get the links 01/12/36= 256 also.

It's not that one method is better than the other at all. It is that in some cases it may be better to look at fractions and ratios. In other cases it may work better to think in terms of a base.

The narrative can even be extended to a single digit number like nine and two digit numbers that end in a base of one. The method in that case is far more cumbersome than just multiplying and linking; so I will save both of us a lot of superfluous energy and skip the explanation of that. I hope you're getting the sense that different narratives can be applied to the same operation. In some cases one narrative may be preferred over another. In other cases there may not be a clear advantage to one narrative over the other. Have the right attitude and try to pick the tool that does the job best for you. That is what a craftsman would do. A craftsman wouldn't try to use the same tool for everything.

Chapter 4: Over/Under 100

We have also seen that a base of one can be used to multiply. We can take a number over a base of 10 and reverse it and multiply easily because of bi-directionality. Now we will use the same process for numbers that are close to a base of 100.

If we wanted to multiply 99 x99, we could doWe have seen how you can multiply numbers that are over or it as follows: 99 is one unit less than 100. Let's put one unit on one finger of the left hand, and one unit on one finger of the right hand. Now, since we are dealing with a deficit from 100. Subtract one of those two numbers from 99 to get 98. Now let's multiply those two ones together to get +01. Here are the numbers we are going to link together: 98/01. Believe it or not 9,801 is our answer.

If we were going to multiply 101x101 the narrative would be the same. He would have one on one finger of one hand and one on the finger of the other hand. This time since we are over a base . We add one to 101 to get 102. now we just tack 1x1 equals one onto 102 to get 10201.

If we were going to multiply 96x99 we would put four fingers out on our left hand, and one finger out on our right hand. We would then either subtract four from 99 to get 95; or subtract one from 96 to get 95. That 95 will be the first part of our answer. The second part of our answer will be the product of the numbers on our fingers. That would be 4x1 equals four. Our complete answer. Then would be 95/04= 9,504.

If we're going to multiply 102x104 we would put two fingers out on our left hand and four fingers out on our right hand. Then we would either add four to 102; or we could add two to 104. Either way, we get 106 as the first part of our answer. Then we multiply together. Then we multiply together the numbers on our hands. That would be 2x4= 08. Our complete answer then would be 10,608.

If we were going to multiply 98x101. We could put two fingers out on our left-hand, and one finger out on our right hand. Now either subtract two from 101, or add 1 to 98. Either way we get 99, and that is the first part of our answer. The second part is the product of the numbers on our fingers. One number is negative, the other one is positive. The product of the two numbers is -2x1= -2. In this case we subtract 2 from 9900 to get 9,898.

This is one of those processes that is so simple that if you done it a few times , it becomes quite easy. The best thing to do at this point would be play with the numbers and see how they work together.

Accordingly, here are some practice problems.

Exercise 3

Problem: Left Right 1st part 2nd part answer

1. -299x99 -01 -01 98 01 9801

2.   97x97 -03 -03 94 09 9409

3.   96x96 -04 -04 92 16 9216

4.   95x95 -05 -05 90 25 9025

5.   94x94 -06 -06 88 36 8836

6.   93x93 -07 -07 86 49 8649

7.   92x92 -08 -08 84 64 8464

8.   91x91 -09 -09 82 81 8281

9.   89x89 -11 -11 78 121 7921

10.   88x88 -12 76 144 7744

11.   87x87 -13 74 169 7569

12.   86x86 -14 72 196 7396

13.   85x85 -15 70 225 7225

14.   101x101 01 102 01 10201

15.   102x102 02 104 04 10404

16.   103x103 03 106 09 10609

17.   104x104 04 108 16 10816

18.   105x105 05 110 25 11025

19.   106x106 06 112 36 11236

20.   107x107 07 114 49 11449

21.   108x108 08 116 64 11664

22.   109x109 09 118 81 11881

23.   111x111 11 122 121 12321

24.   112x112 12 124 144 12544

25.   113x113 13 126 169 12769

26.   114x114 14 14 128 196 12996

27.   115x115 15 15 130 225 13225

28.   102x98 02 100 -04 9996

29.   102x104 04 02 106 08 10608

30.   98x96 -02 -04 08 9408

31.   85x99 -15 -01 15 8415

32.   85x115 -15 15 -225 9775

33.   98x114 -02 14 -28 11172

34.   91x113 -09 13 -117 10283

Chapter 5: Over/Under 20,30...

I bet you're thinking to yourself right now that working over or under a base is very limited proposition. You're limited to numbers that are close to 1,10, or 100. Actually it is not true. With one single adjustment you can work with bases other than 1,10 ,or 100.

For example, if you wanted to multiply 23x31 we could do it by noting that 23 is seven less than a base of 30, and 31 is one more than a base of 30. Now we have an established base, but it's not 10. Let's proceed anyway. Let's put minus seven on the fingers of the left hand, and 100 finger of the right hand. If we subtract seven from 31, or add 1 to 23 either way we get 24.

This is where the other additional step comes in. We simply multiply that 24 by three because our base is 30. 24x3 equals 72, so that is our lead number. No multiplied together. The minus seven and the one to get -7. That is our second number. Now let's put it together 72/-07= 713. Notice how particularly easy this calculation is if you're dealing with a number that is only one unit over a base, as was 31 in this calculation. It is interesting that this problem slightly resembles a synchronized ratio or open carry problem in that the difference between the tens units is one.

If the numbers have the same base the calculation is still similar but without subtraction, and negative numbers. For example, if your multiplying 31x39. You could use 30 is a base too. Now you put one on the finger of your left hand, and nine on the finger of the right hand. If you add 9 to 31 you get 40, or if you add 1 to 39 you get 40. Now he multiply that 40 x 3 to get 120 and that is your lead number.

The back number is the product of one and nine. Our answer is 1209.

If both numbers are closer to but under a base the calculation can be done in a way to assure the smallest possible adjustment to get your final answer.. For example, if you wanted to multiply 39x37. You could make 40 your base. Put -1 on the finger of your left hand. Put -3 on the fingers of your right hand. Take -1 from 37 to get 36. Now multiply that by four. To get an answer of 144. That is your leading number. The trailing number is -1x-3= +03. Your final answer then is 1443. There is another way to solve this one it is an averaging problem also. You could square 38 to get 1444 and subtract one squared from that to get 1443.

This method would also work if you had to multiply numbers that were close a shared base such as 200 300, or 900 for that matter. 991X889 could be done by using a base of 900. One number is 91 units over 900, the other number is 11 units under 900. Subtract 11 from 991 to get 980. Now we multiply that by 9 to get 81/72 = 8820. That will be our lead number. -11x91= -1001. Our final answer is 8820/-1001= 880999. What looked to be a tough calculation wasn't so bad after all.

If the question had been to find the product of 911*989 the work would have been easier your lead number would be ((989+11)x9))= 9000. Now we add to this 11x89= 979. Now the answer becomes 900979. In the exercises we will confine ourselves to two digit numbers with a base other than 10.

Exercise 4

 lead trail answer

57.   21x39 90 -81 819

58.   24x36 90 -36 864

59.   27x33 90 -09 891

60.   87x93 81 -09 8091

61.   54x66 36 -36 3564

62.   73x87 64 -49 6351

63.   71x89 64 -81 6319

64.   82x98 81 -64 8036

65.   52x68 36 -64 3536

Have you noted anything interesting about these numbers? I deliberately picked the first nine because they can be looked at several different ways. In my previous books I called the first ten examples synchronized ratios, or open carry situations. The cross product of the calculation works out to be the smaller number of the calculation. There it is right under your nose.

This happens whenever the units digit sums to 10, and the tens digits are one unit apart. From a ratio standpoint the first problem amounts to adding the two fractions 3/9= 1/3+ 2/1= 2 The total is 7/3, and 7/3x9=21. The shortcut is just to look at the numbers, see the characteristics of the units column and tens columns and just know that the cross product is 21.

The problem is also a averaging problem where each units digit is the tens compliment of the other. The shared base in this example is 30. One number is nine more than that, 39. The other number is nine less than 30, or 21. The numbers on your fingers are equal so their product is the square of nine in this case. That would make the adjustment to 90 be -81. Your answer is becomes 819.

Call them synchronized ratios, open carry, averaging, or over and under a shared base. They all work out the same way. The smaller number is the cross product. There is no need to reinvent the wheel every time you see this situation. The cross product is right there in the open. The toolbox metaphor applies: a craftsman picks the right tool for the job.

The next seven problems are nines compliments, and the next two are eights compliments. The last one is a seven compliment. Again, these were hand picked too. Here is a list of the cross products for the last ten problems compared to what the cross product would have been if the larger number were a little bigger, just big enough in fact to be a tens compliment of the smaller number:

Tens CP: What it was: Difference:

1. 21 19 -2

2. 24 22 -2

3. 27 25 -2

4. 87 79 -8

5. 54 49 -5

6. 73 66 -7

7. 71 64 -7

8. 82 66 -16

9. 52 42 -10

10. 52 37 -15

Notice the pattern? A one unit change in the units sum results in a corresponding change in the real cross product in the amount of the tens cross product times the change amount.

In other words if the total of the units column is nine instead of ten just take that lead number of the cross product and subtract that digit from the tens cross product. This is amounts to a simple adjustment. In the case of problem 11 you take 21 and subtract two from it and you have the right cross product for 21x38.

The rest of the solution is linking together the product of the lead digits and the trailing digits. 3X2/21-2/8x1= 6/19/08= 798.

I digressed off topic quite a bit on this, we will come back to this and all kinds of clever ways to use these properties to a math advantage later on.

Chapter 6: More On Ratios

Previously we talked about multiplying two digit numbers that had a units column that added up to 10, and the tens column was one unit in difference. It turns out that these numbers are over and under a base by an equal amount. For example, if you were to multiply 39x21. You could say that 39 is nine units over a base of 30, and 21 is nine units less than a base of 30.

Because one number is over a base in the other under a base. It turns out that the units are tens compliments of each other, or in other words they had to 10. When this happens , the smaller number is in fact the cross product. In the above example 21 is the cross product, and it's right there under your nose. So you could take a series of problems, like 39x21, 38x22, 37x23, 36x24, 35x25, 34x26, 33x27, 32x28, and 31x29 and easily solve them.

The cross product in each case would be the numbers 21 through 29. You could repeat this process with numbers like 99x81. The same pattern would hold. In fact, for 3 digit multiplication. If you wanted to multiply 624x576. You could do it pretty easily because the cross product would be 576. Then you would have components to put together of 6x5 equals 30, 576, and 1824. The one line component string would be 30/57/1824= 359424. The only difficult part of the whole calculation would be multiplying 76x24.

What I think is even more important to this whole discussion is what happens when you're units digit does not add up to 10. Since we are dealing with tens, the adjustment you need to make to your cross product is very simple. What if you wanted to multiply the 38x 21? In this case your units add up to nine instead of 10. So you are one digit less than a situation where 21 would be your cross product. In this case , you can adjust for being one unit off by subtracting 1/10 of the tens unit unit from 21, so you subtract 2 from 21 to get 19 now you have the correct cross product. If the problem had been to multiply 39x22 you could arrive at the cross product by simply taking 2 and adding it to 22 to get 24 as your cross product.

This proportion holds true no matter what your units total is. Just look at the total, and ask yourself is this over or under ten? If it's over you will be adding your correction. If your total is under 10 you will be subtracting. The amount of the adjustment is that over or under amount multiplied by the tens unit of the smaller number. If you had to solve 87x76 you could do it straight line as 56/97/42=

6612. The cross product being 76+(3x7)= 76+21= 97.

Aliquot or fractional parts can be used too. For example if you were to multiply 53x42. Note that the tens units are one unit apart, and the units digits add up to five which is half of ten. Now your adjustment is ½ of 40, or 20. Your 2 stays tacked onto that so the cross product is 22. Also if the units add to 15 the problem is again an aliquot parts adjustment 1.5x your lower tens number plus the lower units number added back in. For example, 39x26= 06/36/54=1164. The 36 being 1.5x20=30 +06=36.

The next logical question is what happens if the tens units are two units apart. Guess what? The same adjustment is applied, this time using the smaller units digit as the adjustment. For exampre, if you wanted to multiply 49x21. What would be your cross product be? We can go and reinvent the wheel again. If we did this the cross product would be (4x1)+(9x2)=4+18=22. The easier way would be to note that all the conditions are in place to do this calculation as an adjustment to a synchronized ratio. Just add the one in 21 back into 21 to get 22.

I hope you are starting to think some more about this. What if the variance from a perfect open carry was one unit each? Well, let's look at that scenario. Let's say you wanted to multiply 48x23. The tens units add up to 11, so that is one unit more than ten. The tens units are 2 digits apart and that is one digit more than we would have in a perfect open carry or synchronized ratio situation. The adjustment in this case is equal to the sum of the digits in the smaller number being added to the smaller number. In this case we add 5 to 23 to get 28. Let's check that using vertical and crosswise. $(4\times3)+(8\times2)=$ 12+16=28.

If one number were over and the other number was under then the adjustment would be the difference between the digits. For example 46x23, the components being 08/23+1/18= 1058. This one is a little tricky because 2-3=-1 but our adjustment is positive because 3-2=+1. You can expand this adjustment to any combination of two digit numbers you want. Let's look at 56x78. Here we are 4 units over 10, and one unit more than the required one digit. Our correction to 56 would be what? It would be (+1x11)+ (3x5)=26. The cross product is 82. You can get it by adding 1x6 to 4x5, or by taking the sum of the digits times one and adding 3x5 to that. Either way you get 26 to add to 56 to get 82.

The single digit multiplication model of this form is instructional and may be a good place to develop some confidence in the method. Let's say you wanted multiply 73x9. The units digits add up to 12, when the tens digits are six units from the required one digit difference. The first vertical multiplication is zero since zero times seven is zero. The cross product is 6x9+9 equals 63. The final vertical calculation is 27. Now we had the components 00/54+9/27= 63/27= 657

I'll be the first to admit that this is pretty lame, but I want to make the point that the narrative can even be applied to single digit numbers. To do the full multiplication you just assemble three components, one which happens to be zero. The second component of the cross product in this case is always half done for you in that the first part of it is zero.

An example with 11 as the smaller number may also be instructional. Let's say you wanted to multiply 74x11 In this case the tens digit is five units greater than the one we look for. The units digits add up to five, so that is five units less than what we need. Our cross product is ½ of 10+ 01= 06., plus the original 5x1 for a total of 11. The shortcut on this is that anything with a digit sum of 11 and two digits multiplied by 11 has a cross product of 11. Try it with 83x11, 92x11, 29x11 ect.

Anything with two digits you multiply by 11 has a cross product equal to the sum of the digits. That is why they say to add the neighbor to multiply by 11. 96x11 then would have a cross product of 15. You could get it by subtracting 3 from11 to get 8. Then add back in 7x1=7. The total is 15. The alternative would be to subtract 7 from 3 to get 4 and add that to the 11 to get 15.

Chapter 7: Squaring Two Digits In Two Steps

You can square any two digit number by taking the sum of the units column and then carrying forward that amount to the larger tens digit. Then multiply that amount by the smaller tens digit. That result is equal to the first vertical calculation and the sum of your cross products. The final step would be to the last vertical calculation. It sounds more complicated than it is.

Almost everyone knows that you can square a number that ends in five by multiplying the tens digit by one unit more than its value and tacking on 25. For example, if you wanted to square 25. You could multiply 2 x 3 to get 6. The next step would be to tack on 25. Your final result would be 625. In a similar fashion. 35 squared would be three times for equals 12, and tack on 25 to get 1225. You could do the same calculation for any two digit number that ends in five. For the numbers 15, 25, 35, 45, 55, 65, 75, 85, 95.

If you think about it the calculation we are doing is exactly what we described in the very first paragraph. Since the total of the units digits is 10, we are carrying forward, 10 to the tens column. That then becomes a one to be added to one of the two numbers. You could do the same thing no matter what the units add to. If you wanted to square 11, you could add the units digits to get 02. Now carry that forward to the tens column and add to one of the tens. One would become 1.2x10. So now we are working with 12 and we are multiplying it by the remaining one in the tens column. That would give us our first vertical calculation and the sum of the cross products all in one step. All that remains now is to link up the product of the units column which is one times one.

Our final result becomes 121. We just squared 11 in two steps.

Let's try another example. Let's square 24. The units add up to eight. So we are taking 28 and doubling it to get 56. We know that 4x4=16. If we link 56/16 we will have our final answer of 576. I hope that you are brimming with ideas about squaring numbers at this point. Some numbers in particular would be especially adaptive to this method. Anything that began with a one for example would be absolutely easy. Once you added the carry forward amount you have your cross product right under your nose. Let's square 17. 7+7 equals 14. Add ten to it to get 24. To that we have to link 7x7, which is 49. Our final link would be 24/49= 289.

In a similar fashion, 19 squared would be 28/81= 361. Squaring anything in the 20s would be particularly easy also. We are just doubling the sum of our carry forward amount and one of the tens digits. For example, 28 squared would be 36x2=72. To this we would link 8x8= 64. Our final result would be 72/64=784. 29 squared would become 76/81= 841. Numbers in the 30s, 40s, and 50s would all be particularly easy.

In that two digit squares are so useful. I generally recommend that you just memorize them if you really want to become mental math whiz. This method would be close second to rote memorization. In one of my earlier books I had a long section about the importance of being able to multiply, double and triple digit numbers by a single digit. I hope this demonstration of carrying forward and doing double-digit by single digit multiplication shows you the power of these two items. The method extends a shortcut that everyone knows to all double digit numbers that you would want to square. You don't need no stinking calculator to square a two digit number!

One often overlooked fact is that all of these methods are bidirectional. For example, let's look at 25 squared again. We could reverse the way we look at the columns and proceed from there. In this case then you add the sum the tens column and carry it backwards. That would give you the number 45 in the units column. Then you would multiply that by the remaining five in the units column. That would give you 225. To that you would add the product of the tens column which is 2x2 equals four. The four would go on the front end of the 225, in the form of 04/225= 625.

In a similar fashion, you could square 35. You would add the threes together to get six. Bring it over to the units column to get 65. Multiply that by five to get 325. Then link it with 3x3. Your final result would be 09/325= 1225.

I hope you realize that this is a bit like Tesla and Edison arguing over AC and DC current. AC current is, of course, bidirectional. DC current is one directional. The big argument back then was which was better. Realizing the advantages of bi-directionality is sort of like realizing the advantages of AC current. You can transmit your leverage or power so much farther by using bi-directionality. We mentioned how effective this technique is for squaring numbers in the 10 through 50 category. You can double that effectiveness now through bi-directionality to apply to squares that end in the numbers one through five.

For example, if you want to square 92, you might want to consider adding the tens columns and carrying that sum backwards to one of the twos. That would give you

182. Now just double that number to get 364. To that link the product of 9x9. The final result would be 81/364= 8464. That might be easier than multiplying 9x94=846. With bi-directionality you have an option to work the smartest way. You don't have an option to work smart with the grade school method you have been stuck with the majority of your math life. The option to work smart is huge. In the three digit world you have options about how to split the number, and now even an option that allows you to work after the split with the easiest numbers. How great is that?

The best way to use bi-directionality is to look at the number your squaring and choose the column with the smaller digit in it as the column that you will use to multiply by. The column with the larger number would always be the column that you sum and carry either backward or forward to. In this way you would always have to multiply by the smallest number possible. It's easier to add the larger numbers, and multiply by the smaller number than vice versa. For example, if you wanted to square 19 you would carry forward. Conversely, if you wanted to square 91, you would add the nines and carry backwards.

Exercises:

Square these numbers mentally, just write down your answers. Work them according to number order so that you are taking advantage of bi-directionality and get to experience it on real problems.

1. 13

2. 14

3. 22

4. 23

5. 32

6. 36

7. 41

8. 49

9. 38

10. 52

11. 57

12. 63

13. 69

14. 74

15. 78

16. 82

17. 86

18. 92

Chapter 8: Squaring Three Digits In Two Steps

You can square any three digit number in a similar fashion to how we did it for two digit numbers. This time you split the three digits into a one digit number and a two digit number. The choice of putting the two digits first or last is entirely up to you. An example might be the best way to give you some ideas about how you may decide to split a three digit number you wish to square. If you wanted to square 124 you could split it into 1/24, or 12/4. Which would you choose? Any number that starts or ends with 1 is pretty simple to square just using the base and overage method. I know the answer to this one immediately. You are 24 units over a base of 100, therefore add 24 to 124 to get 148. Now link that to the square of the amount of the overage. 24 squared is 576. Our final answer is 148/576=15376.

The generally accepted easiest way to do this one without the base and overage narrative is exactly the same. Split 124 into 1/24. The units add up to 48 so we carry that forward to the 100. Our new number to be multiplied by 1 is 148. 148x1 is 148, that is the first part of the solution. I know my two digit squares so the rest is quite easy, just link 148 and 576 to get 15376. The other approach might be to split this into 12/4. I know that 12x12=144 so I can proceed to use vertical and crosswise. From a fractions standpoint 12/4= 3. If I double that I know my middle component is 6 times 4x4. The links would involve putting together 144/96/16= 15376. I can also carry back the sum of 12+12. The multiply 244x4 to get 976. This would result in assembling 144/976=15376. We just solved this five different ways. We used vertical and crosswise, fractions, carry forward, carry backward, and over a base. They all worked but the easiest for me at least was to treat this as a over a base of 100 problem, or a carry forward with a split of 1/24 so I could use the power of the identity property and just link two components 148, and 576. The important point is that just seeing one as the first digit immediately tells you the easiest way to solve this problem.

Another example might involve just turning the number around and squaring it. Let's look at squaring 421. In this case you could split the number into 4/21, or 42/1. I know 42x42=1764, or at the very least I could get that part by using a fractional approach. 4/2x2=4, so my components to link are 16/16/4=1764. Now I just link 1764/841= 177241. Again, once you see the one as a digit that is the number you want to gravitate towards. The base narrative yields the exact same result. 421 is 420 units over a base of 1.

Add that overage to 421 to get 841. Now link that with the product of 42x42.

I always split my three digit squares so that the single digit is the smaller of the lead or trailing digit. In other words I split so that my two digit square is as big as I can get. It is consistent method, so I don't lose a lot of time kicking the tires so to speak. The procedure should be to carry back or forward towards the smallest number of the two digits.

I hope this gives you a feel for squaring three digit numbers. Every multiplication is different but at least you have a guideline for doing the work. The way you really get good at this is to work with the numbers. Remember the 10,000 hour rule. It takes a long time to get really expert at anything. You can get relatively proficient very rapidly though. That part of the journey will be extremely rewarding and may be the most fun.

Exercises:

Square these 3 digit numbers. Show how you would split them, the two step components, and the final answer.

1. 789

2. 456

3. 123

4. 741

5. 987

6. 654

7. 321

8. 147

9. 758

10. 367

11. 532

12. 235

13. 498

14. 894

15. 265

Solutions:

split into: Components: answers:

1. 7/89 6146/7921 622521

2. 4/56 2048/3136 207936

3. 1/23 146/529 15129

4. 74/1 5476/1481 549081

5. 98/7 9604/13769 974169

6. 65/4 4225/5216 427716

7. 32/1 1024/641 103041

8. 1/47 194/2209 21609

9. 7/58 5712/3364 574564

Chapter 9: Use Aliquot Parts

Just as you can square numbers with the sum and difference method, you can also multiply numbers that are not identical. Let's look at a commonly used example of mine. We want to multiply 39x21. We know that since the units column totals to 10, and there is a difference of one between the two digits of the tens column that the cross product is in fact 21. No multiplication necessary. From there, you can quickly figure that 39x21 = 819.

If we didn't know all that we could still derive the answer by noting that the units column sums to 10. Then we would carry forward the 10 to the tens column where it would become 01 to be added to the larger number three. Then we would multiply (3+1)x2= 8. Now we have the first digit of the solution. For the second digit we could note that the difference between the original numbers in the tens column is one. Now we multiply that one by the one in 21. Or to put it another way, we multiply that difference of one by the units digits of the smaller number. In this case one times one equals one. That number is our second digit . We are two thirds of the way to our answer. The final digit is merely 9x1, or the vertical product units digits. This same approach works for all two digit calculations.

If we wanted to multiply 78x32 we could do it by noting that the 2 and 8 are tens complements. Now we add one to the 7 in 78 and multiply that (7+1) by 3 to get 24. The second step is to find the difference between seven and three which four. Then we multiply that 4 by the 2 in 32. That gives us 08. Our final component is 8x2= 16. Let's put those components together; we have 24/08/16= 2496.

In fact, regardless of what the units digits add up to we can still use the sum and difference method to get an answer. If we wanted to multiply 79x33. We could do it using the same method. Since 9+3 equals 12, we will now carry forward to the seven in 79. That gives us 82. Then we take 82, and multiply that by 3 to get 246. The difference between the tens digits is still 7-3= 4, and we multiply that difference by the 3 in the units column of 33. Our second component is 12. The final component is 27, let's link them together to get 246/12/27= 2607. This may not seem like much of an improvement over the standard vertical and crosswise method. find that it really does help the calculation flow because carrying forward automatically links some of the digits together. You don't do a calculation set it down, do another one set it down. The calculation melds together a little better using this approach. The hardest part of mental math once you get the hang of it is keeping track of the digits and components as you calculate. Anything that facilitates that is extremely useful.

I'll show you another reason why this method has some advantages. Let's change the example we just did slightly. Let's say we want to calculate the product of 79x35. Now the units digits add up to 14, and we carry just the 10 part of the 14 over to the 7 in 79, 7+1= 8 which we need to multiply by 3. That calculation gives us 24. Now we take the difference between the seven and the three which is four. Note that we also have a difference of four from when we left 4 on the units side earlier. Our cross product is going to be 4 times the sum of the digits in 35. That means, our cross product is 4x(3+5)= 32. Our components become 24/32/45= 2765.

Using this little trick with the difference equaling the sum works in all kinds of situations you don't normally recognize. For example, if you wanted to multiply

125x7 you could easily do it by recognizing that 5+7=12, and 12-0 also equals 12. From there just link 12x7 with 5x7. The components would be 84/35= 875. I don't even think of this process as multiplication, but in fact it is.

If we expand this process to a 3 digit by 2 digit multiplication, an example might be 199x19. This has lots of nines in it and appears to be a little onerous to do. Let's see what we can do with it. The nines add up to 18, and the difference between 19 and 1 is also 18. Our components become 19/180/81= 3781. That middle component is 18x(9+1)=180.

There are any number of examples where this works. We can use it on mixed numbers of digits or on problems where the numbers of digits are the same. In some cases where the digits are the same we have the added step of taking the extra digit over to the tens column. in situations where the units sum of the digits is less than ten we can dive right in and use the method straight away. If we multiply 98x29 we would make the 9 a 10 since 8+9=17. Our first component would be 20. After that the 7 in 17 and the difference of 7 in 9-2 tells us the second component would be 7x(2+9)=77. The final result would be 20/77/72= 2842. If we wanted to get 96x21, we could just write down the components as 18/21/06=2016. The calculation couldn't be much easier.

The sum and difference method has some of the features of the open carry method we discussed in a previous book. For example, if we wanted to multiply 57 x 42, we could do it by multiplying 59 x 4 for a first component. The second component would be the difference between the digits in 5 and 4 multiplied by the two in 42. The assembled components would then be 236/02/14= 2394. If we turn the problem around and went to multiply 57x24. We could do it by adding one to the five in 57 and multiplying by two to get 12. The next component would be the one that we left behind times the two in 24 + (5-2) times the four in 24. That would give us 2+12 equals 14 for second component.

The method lends itself to some shortcuts that are not available in the traditional vertical and crosswise system. There are some other nice features as well. One being that the method is subject to aliquot parts. For example anything that you had to multiply where the units add up to five would count as one half when you took it over to the other side. For example let's multiply 84x21. In this case the units add up to five. Our first component would be 85x2 or 17. The second component would be 06, and the final component would be 04. The solution would become 17/06/04= 1764. We will see that aliquot parts become really important in the three digit world.

Notice that if we were multiplying 84x31 or 83x32 the units would add to 5 and the difference between the numbers would be five too. Now the components can be assembled quite easily. The first component would be the vertical calculation of 8x3=24. The second component would be 5 x (the sum of the digits). In the one case 5x4=20, in the other 5x5=25. The final component would be the vertical calculation of the units digits for each problem. In one case we have 24/20/04=2604; in the second case we would have 24/25/06=2656.

If the base is the same in each case, in either the tens column or the units column the multiplication is a simple two step process not much different than squaring. If you wanted to calculate 39x38 just multiply 47x3=141, and 72 to it to get

141/72=1482. By the same token 93x83 could be done by adding 17 in front of the 3 in 93. Multiply that by the other 3 to get 519. Now link 72 in front of that 72/519= 7719. It is a two step process just like squaring is.

One other easy Aliquot part is when the digits add up to 15. The amount you are carrying forward or back is 1.5 or 15. That works out nicely when multiplying. If you wanted to multiply 97x98 you could multiply 105x9 to get 945, now link that with 8x7=56. Your final result would be 9506.

Remember the fact of bi-directionality also. In the case of 57x42, we could note that the tens column adds up to nine. Now we carry that nine in front of the seven in 57 and double it. That would give us 194. To that we add in the difference of 5 in the units column multiplied by 4. The adjustment is 20 and 194 is changed to

394. The other link is 5x4=20. Our final result is still 2394.

The same rule would apply that you used when squaring. That would be to carry forward or backwards towards the smaller digit in the smaller number as much as you can. Unless the sum or difference from 10 is exactly the same on both columns don't try to get too clever with the cross products. Let's try some exercises and see how it goes.

Exercises

Multiply these two digit numbers in your head just record your answers on paper as you go along.

1. 98x78

2. 87x98

3. 54x47

4. 65x57

5. 52x69

6. 74x86

7. 47x96

8. 67x49

9. 32x58

10. 21x91

11. 64x36

12. 23x94

13. 78x91

14. 48x72

15. 97x52

Answers

1. 98x78= 7644

2. 87x98= 8526

3. 54x47= 2538

4. 65x57= 3705

5. 52x69= 3588

6. 74x86= 6364

7. 47x96= 4512

8. 67x49= 3283

9. 32x58= 1856

10. 21x91= 1911

11. 64x36= 2304

12. 23x94= 2162

13. 78x91= 7098

14. 48x72= 3456

15. 97x52= 5044

Chapter 10: Three Digits Any Numbers

Three digit multiplication of numbers that are not squares is also possible using a sum and difference method. The easiest problem type after doing squares is numbers that contain partial squares. For example, if you had to multiply 315x215. The split would logically be 3/15, and 2/15. Clearly you could do this as a fraction. The sum of this fraction is 5/15ths, which is the same as 1/3rd. If 15x15=225, the middle component is 1/3rd of that or 75. Now just assemble 06/075/225= 67725.

You could do the same problem in three steps with the sum and difference method. Add the 15s and carry forward to the 3. Then multiply 33x2=66. Take the difference between 3 and 2 and multiply that by 15. Finally square 15 to get the last component. The assembly would be 66/15/225= 67725. I don't know that one method really has an advantage in this case, but having an additional way to check your work that is independent of the first method is a huge advantage in terms of accuracy.

The secret about splitting the three digits so that you have a two digit part that adds up to 30,40,50, and so forth is just to find a combination where the units or tens column is comprised of tens compliments. For example, if you were to multiply 852x728 you can immediately see that you have tens compliments in the units column. A closer look shows that the last two digits total up to 80. The math after that is quite simple. The components become 616/28/1456= 620256.

If the problem were changed so that the middle column had the tens compliments you could solve it by making the split between the digits after the first two columns. For example, you could solve 528x287 easily by splitting it into 52/8 and

28/7. We know that 52x28=1456, or we can easily calculate that it is. Our components become 1456/28/5656=151536. The 28 comes from (8-7)x28. The 5656 comes from 52+28=80, added to the front of the 8 in 528 and multiplied by the 7 in 287.

You can also have aliquot parts with three digit multiplication. The parts that are especially useful are situations where the two digit column adds to 50,75,100, and 150. You always want to take advantage of a situation where the columns after you split them are 1, 5, 10, or 15 units apart as they are all so easy to work with when you compute components. For example 926x449. Split this into 9/26 and 4/49. Since 49+26=75 you will end up multiplying 9&3/4x4 to get 39 as your first component. The second component is 5x49, which is an easy aliquot multiplication. Half of 490 is 245 for your second component. All that is left is 26x49=1274. Once you have the components the rest is 39/245/1274=415774.

If the numbers do not afford you a easy answer, you can still use the method. Just split you digits towards the smallest digit of the smaller number as a general rule. For example if you wanted to multiply 895x349 how would you proceed? I would split the numbers into 8/95 and 3/49. Then I would multiply 944x3 to get 2832 as the first component. The second component would be 5x49=245. Now we have 3077 to link with 95x49. That would be 3077/4655=312355. It looked worse than it was, you could take it in smaller parts after you got your 3077. Just link 3077/(104x4)=31186. Now link 31186 with 45/45 or 495. That link would be 31186/495=312355. Break the numbers into small parts and just keep building to get your answer.

Recall when we did two digit numbers that if the difference between the tens digits was the same as the units digit portion of the sum of the units digits, then our middle component became the sum of the digits in the smaller number multiplied by that difference. For example 41x12 has a difference of three in the tens column, and the units digits add up to three. We could do this one straightaway as 4/(3x(1+2))/02= 492.

The same thing occurs in three digit multiplication. For example let's multiply 724 x 226. We can see that the difference between the 7 and 2 is five. We can also see that 24+26 equals 50. The math after that is quite simple. Our components are 14/(20+26)x5))/624. That would give us a final answer of 14/230/624= 163,624. The middle component being 2(0) added to the number 26 to get 46, which we then multiply by 5 to get 230.

If we wanted to multiply 311x109. We could do it in a similar fashion. In this case, the difference is two, and the sum is 20. Now we just proceed to put the components together. That would be 03/38/99= 33,899. You can also use this in a aliquot parts fashion. For example, if we change the problem we just did, so that we were multiplying 211x109. In this case, the difference is one in the hundreds column. The units column adds up to twice that though. Now our components become 02/29/99= 22,999. The middle component became (2x10)+9. Note that it did not become what you would presume to be a logical one half of 38. If you play with this one a little bit you will get a good feel for how it works, and it may be useful to extend the range of this little trick.

The same technique can be used in mixed digit multiplications such as 315x15. In this case the 15s add up to 30 and 3-0= 3. From here we proceed. There is a first component, it just happens to be 3x0=0. The second component becomes 3x(15)=45, or 30+15=45, to keep it in digit sum terms. To this just link 225 to get 4725. We could have done this same thing with any combination that added up to 30 involving the sum of the last two digits of the multiplier and multiplicand.

For example 314x16, 313x17, 312x18, and so forth would all have zero as the first component. The second component would only change slightly it would be 3x16, 3x17, and 3x18 in that order. The last digit would be 14x16, 13x17, and 12x18 to complete the multiplications of these 3 numbers.

There are a lot of other combinations where this works. Three digits just exponentially increases the opportunities to apply this. Then there is the situation where the "units digits" adds up to a multiple of 11, and the hundreds digits are separated by that sum divided by 11. For example, let's look at 829x326. The split on this would be 8/29 and 3/26. It hardly lends itself to using fractions to find the components. Since 29+26=55, and 8-3=5 we can solve this much the same as we would a problem that summed to 50 and had a difference of 5 in the hundreds column. Our components would become 24/59x5/29x26. The 59 deriving from adding the 3 in 326 to each digit of the number 26. Our final result would be 24/295/754=270254.

Another easy situation occurs when the units digits add up to 11 and the tens add up to the same number that the hundreds digits are separated by. The problem of 829x322 exemplifies this situation. The numbers 29+22= 51, and the hundreds digits are 5 units apart. Can you guess what the cross product will be? The components of this one are 24/263/638= 266938. The 263 comes from adding 30 to the 22 to get 52. Now you multiply that by 5 to get 260. The three is tacked on after that with no additional multiplications to it.

Let's do some exercises to increase our confidence and proficiency with three digit multiplications.

Exercises

1. 219x48

2. 999x819

3. 924x336

4. 518x302

5. 242x818

6. 511x219

7. 787x638

Answers

1. (21/9)x(4/8)= 84/204/72= 10512

2.   (99/9)x(81/9)=   8019/18x(81+09)/81= 818181

3. (9/24)x(3/36)= 27/396/864= 310464

4. (5/18)x(3/02)= 15/64/36= 156436

5. (2/42)x(8/18)= 16/372/756= 197956

6. (5/11)x(2/19)= 10/117/209= 111909

7. (7/87)x(6/38)= 4950/38/3306= 502106

Chapter 11: Solve For Something Else

Sometimes the best way to solve a math problem is to solve for something else. This is not an uncommon problem solving technique. It's done all the time in all sorts of applications not related to math. It is a great technique to learn and then use elsewhere as a life long problem solving skill. I can't think of a better subject to apply this thought process to than math. Let's go to a simple example that demonstrates the principle involved. If we wanted to multiply 18x68, I hope your immediate response would be to think in terms of fractions. You have 1/8th, and 6/8ths. The two fractions add to 7/8ths. You know your components to be 6/56/64. The answer is 1224.

Another approach that you could use would be to say to yourself that with the addition of 1/8th the fraction would become 8/8th, or the whole number one.

The middle component would be the same as the last component or 8x8=64. To get there you could change 18 to 28, or multiply 1x8, and subtract that from 64 to get 56. In this case we took an easy problem and made it more left handed. It does not always work out that way though. If the problem had been to solve for 116x78 I would still think fraction. This time if you added 1/16th to 1/16 you would make the second and third components equal at 8x16=128 each. This time just adjust slightly and subtract 8 from 128 to get 120 as your second component. So we solved initially for a second component of 128 and subtracted 8 to get 120.

The salient point is that the correct adjustment of one unit on one side of the split numbers was to adjust the diagonal number from it by the value of that number. In both these examples that adjustment was 8, and it was subtractive. The adjustment would be additive if the addition of one more unit got us to a result that was almost instant. We could extend this out another increment by just doubling our adjustment too. All this should not be surprising considering our experience with open carry methods.

Under that system you manipulate your adjustments similarly. For example, if we wanted to find 38x21 we could take a similar approach. We know that if we had to find the cross product of 39x21 it would be 21 right under our nose as the units add up to 10, and the tens are one unit apart. What would it take to adjust that to get the cross product of 38x21? The adjustment needs to be subtractive. It should be the value of the tens unit in the smaller number. So we adjust the 21 downward by 2 units. The cross product is 19.

If the question had been to solve for 47x32 we could easily do this one as 12/(32-3)/14=1504. If we were trying to solve for 48x34 the adjustment would be for 2 units from 34 and would be additive. The components would become 12/(6+34)/32= 1632. We have applied the occasion to solve a problem by solving for something else to fractional situations and to open carry situations. Where else would they be applicable?

In a previous book we covered numbers that were close to a base of one, 10, 100, and even 1000. I hope you've asked yourself the question, "What would the adjustment be if we were working in that system?". For example, if we wanted to multiply 115x115 we could do it by noting that each number is 15 units over a base of 100. Then we would add 15 to one of 115s. That would increase it to 130. To that we link the product of 15x15. The final result would be 130/225= 13,225. What if we were presented with the problem of finding the product of 215x115? When you're first presented with this problem, you might look at it and think to yourself that if only that 2 in 215 was a 1, you would have this one knocked out in no time. The truth is that you can do that by taking the prefix of 115x115, the 130 and adding 115 to it to get 245. Now link in the 225 to get 24725. If the question had been to solve for 315x115 you could do it in a similar manner. This time just add 115x2=230 to 130 to get 36/225=36225. In this case it is easier to get this answer than it is to solve for 215x115.

Remember that the last two digits don't have to be identical, they were in this example only. You could solve for 112x213 by adding 112 to 125 to get 237. Then you would link the product of 12 and 13, which is 156. The final result would be 137/156= 23,856. 313x112 would result in 35056.

If you were looking at multiplying 313x412 . You could start off with the prefix for 313x312. That would be 325x3= 972. To that add 313 to get 1288/156=128956. What we really did with all these little fixes is come up with a way to get the easy answer, and then adjust our way easily into an answer for a different problem. We have applied this method to problems involving fractions, problems involving the identity property, and problems that are close to being an open carry situation. There is yet another way, we can apply the same principle.

The method in the case of base related problems is especially powerful. You can apply it to a mixed digit calculation such as 107x54. The answer would become (5x54)+(5x61)/28=270+305/28=5778. The 5x54 is the difference between 10&5

multiplied by 54. The adjustment is 5x(54+7)=305. The 28 is from multiplying 7x4. 324x54 would work out to be (32-5)x54=1458+(58x5)=1748/16=17496. The 58 would be the result of adding 54&4, the multiplier 5 is from 54 having a base of 5(0).

Another simple example of solving for one thing to get the answer to something else comes along in the form of problems that are close to the shortcuts we are all familiar with. For example, we had to solve the problem of 16x15. We could first solve for 15x15. You will recall that this is a real easy multiplication. The answer is 225. What do we have to do to correct 225 to get 16x15? All we have to do, it turns out is add another 15; and that is considerably easier than doing the actual multiplication. Could we do 14 x 15 in a similar fashion? The answer is yes, in this case we would square 15, and subtract 15. The result would look like this, 225-15=210.

You could also apply this incremental approach to multiplications involving reciprocals. For example, 27x72 involves multiplying a number by its reciprocal. With reciprocals first coefficient and the last coefficient are identical. In this case, they're both 14. The middle coefficient is always going to be the square of the two digits added together. The middle coefficient in this case is (2x2)+(7x7)= 53. If you multiply 27x73, hopefully you would observe that this is within one number of being a reciprocal. One way you could solve it would be to take 1944, and add 27 to it. That would give you 1971. Even better would be to take that additive change of 1 and add 1x2 to it. In this case the adjustment is again the amount of change multiplied by it's diagonal opposite, which is 2. So you add 20 to the cross product, and in the third calculation put in 21 instead of 14.

There are other ways that you could use this incremental approach. We know the number 11 has some unique properties. For one thing, the sum of the odd coefficients equals the middle coefficient any time you multiply a number by 11 or a multiple of 11. You may occasionally look at a problem and notice that you are almost multiplying by multiple of 11. Let's take the example of 23x64. That 23 is one unit away from being a multiple of 11. Let's go ahead and multiply 64 x 23 and treat the middle component for a moment as though it were the middle component of 64x22. We start out with a first component of 6x2=12. This calculation is the same for either problem. The 2nd component of 64x22 is real easy to get it is 2x(6+4)=20. Its just a single multiplication of the sum of the digits in 64. We know the cross product of 64x23= ((6x3)+(4x2))=26. The correction to the 20 has to be to add 6 units. You get that by multiplying the change of one unit by it's diagonal opposite which is 6. The final component is going to be 4x3=12, instead of the 8 you would get from 2x4=8 in 64x22. Our final answer on this one would be 12/26/12=1472.

So now we have a whole new method based on the unique properties of 11 and multiples of it. You could use this one anytime you see numbers that are close to being a multiple of 11. You can use it in a row situation, or you can use it in a column situation. It is also bidirectional, so it has enormous potential to be a very effective tool. The method would be especially effective when the adjustment is only one unit because your correction, then is always a single digit. Just for fun, let's look at a problem like 98x97. We already know we can solve this using a under a base method. 98 is two units under 100, 97 is three units under 100. Now we subtract three from 98, or subtract two from 97, either way, we get 95. To this we tack on 3x2=06. The easy final answer to this is 9506. We could check our work by multiplying 9x9 to get 81, and then we could link it with (16x9)-9=135. Now we have to put together 81/135/56. That gives us 9506. This approach was not too much more difficult than the first method we used solve a problem.

I deliberately selected this problem to make another point. If you think about it for a moment you already had multiple of 11 in the first column. In this case, you could of written your components straight across as 81/9x(8+7)/56= 81/135/56= 9506.

I am trying to make the point that using a multiple of 11 in a column is quite easy since you multiply your components in a column fashion as well. If you don't look at the problem in both a row or column format, you may miss the obvious. The further point is that having the same base in either the tens column or the units column presents the problem as a multiple of 11 situation. Sometimes we get so caught up in the narrative of having the same base that we miss other interpretations that can be quite insightful. In this situation though remember that the first and last coefficients will not add up to the middle coefficient, the horizontal calculations will add up to the cross product.

Chapter 12: Multiplication By 9

a) Multiplying the same number of 9

with the same number of digits. Eg: 999 *
abc

So, the first step in multiplying abc

number digit with the same number of 9
is done by:

abc * 999

= abc * (1000 - 1)

= abc000-abc

= a/b/(c-1)/(9-a)/(9-b)/(10-c).

In short, the last three digits while

multiplying 999 * abc, is 1000 – abc.

b)Multiplying with the more digited

number of 9. Eg: 9999 * abc

If there's more numbers of 9, for

example: multiplying abc with 9999,

take abc as 0abc and follow the steps in the first case.

abc * 9999

= 0abc * (10000 - 1)

= 0/a/b/c/ (9 – 0)/ (9 - a)/ (9 - b)/ (10 - c)

Continued to next page

c) Multiplying with the less digited number of 9. Eg: 99 * abc.

Take the above example,

abc * 99

= a: {bc – (a + 1)}: (100 – bc)

You put the colon marks after the number of 9 digits and so, we are taking two digits at a time. In the middle (the two digits at a time. In the middle (the subtracting term), you write the two left numbers and add 1 to it and then subtract it.

Eg: 56789 * 99

= 5: 67 – 5: 89 - (67 + 1): 100 - 89

= 5:62:21:11

=5622111

REMAINDER

The remainder of any number after it is divisible by 9 is very easy and here the concept of Navasesh comes in. Suppose you are asked to tell the remainder of abcd after division with 9. This is easily done by a+b+c+d and if that results in two-digit xy then again sum it up until you get one digit less than nine. And that's all.

For eg: remainder of 6978 after dividing by 9 = (6 + 9 + 7 + 8)

= 30 (3 + 0) 3.

Therefore, after dividing 6978 with 9 we would get 3 as REMAINDER.

And for Navasesh of 6978, it is shown by N(6978) = 3, which got earlier by summing up everything.

You can take out the integers which add up to 9 or actually are 9 to improve your speed. Therefore, in this example, we can just add

6 + 7 + 8 and then we would get 21, which again after summing up 2 + 1 gives 3. Therefore, that's it.

MULTIPLICATION

So, when you multiply a number xy by 11. The resultant number takes form of x/(x + y)/y. It's that simple. For larger numbers, like multiplying 1423 * 11 = 1/(1 + 4)/(4 + 2)/(2 + 3)/3

= 1/5/6/5/3

= 15653.

For multiplying numbers like 111, 1111, etc. the same process goes, but the numbers added is equal to the number of 1's.

For eg: 5967 * 1111

= 5/(5 + 9)/(5 + 9 + 6)/ (5 + 9 + 6 + 7)/(9 + 6 + 7)/(6 + 7)/7

= 5/14/20/27/22/13/5

= 6629337

That's all.

Chapter 13: Multiplication With Bases (Nikhilam)

1) PRIMARY BASE.

For this, we use a Primary Base (bp). Sometimes, we use two bases – a

Primary Base (bp) and Secondary Base (bs). Okay, let's get it done with an

example. We'll go to Secondary base in the next part.

Let's take a b * x y. And let a and b
x y are two numbers close to 100
. Now, as they both are close to 1
00, therefore, first we would w
rite both of them one above oth
er on the left side and on the rig
ht side we would write their dif
ference from 100. Now, we wou
ld cross-add or cross-subtract
as per required. And

write the first result down.

On the right side we would mult iply and write normally i.e., if t he result of (ab−100) is a negati ve number then we would write a negative number. And then m ultiply the results on the right up and down with each other. A nd also write the result separat ed by '/'.

Let the result of cross-additio n be Λ and the result of the multi plication

with subtractions from 100 be B.11

Now, we carry out normal subtraction or addition as per required. This is

determined by whether the multiplication of the right side is negative or positive.

Therefore,

ab * xy = A/+-B = A/B or (A - 1)/(100-B)

Now, real example: 98 * 103

Here, our base is 100.

98 - 100 = (-2); 103 - 100 = 3;

98 (-2)

103 3

101 / (-6)

Here, by both cross-addition,

we get the same result: 98 + 3 =101 and also 103 + (-2) = 101

And on the right side (-2) * 3 = (-6)

Now we would carry out normal activities 101/(-6) = 10094

Therefore, 98 * 103 = 10094.

TRYTHESE :

a)97*101,

b)105*104,

c)101*109

1 2ANSWERS:a)9797,b)10920,c)11009

2) SECONDARY BASES.

Now, we take in the case of bs. Generally, bp are 10, 100, 1000, etc. And bs are anything other than 10, 100... like 50, 70, 250, 600, etc. The overall steps are same. We just add another trick for taking the factor out of the bp. What I mean, is if suppose the bp is 100, and the bs is 50, then Factor = bs/ bp. Here, Factor = 50/100 = ½

One thing I would mention here, is that we could take the bp as per our choice and conveniently. Therefore, we took the bp as 10 in the above case, then, the factor would have been 5. Next Example: 56*53 Here, we would take bp as 100 and bs as 50.

Therefore, the factor is ½.

5 6

5 3

5 9 / + 1 8

= 5 9 * ½ / + 1 8

= 2 9 . 5 / + 1 8

= 2 9 / 5 0 + 1 8 = 2 9 / 6 8 = 2 9 6 8 .

One thing, I would mention here is that the result WOULD BE SAME IF WE USE THE bp AS 10. We would need to multiply 59 by 5 and carry 1 in 18 i.e., 59 / +18 = (59 * 5)+1/8 = 296/8 = 2968.

That's all you can also do the same with other bases like 250, 700, etc.

That's all.

T R Y T H E S E :

a) 256 * 249 (take b s as 250 and b p as 1000, hence, factor is 1/4.)

b) 789 * 801 (take b s as 800 and b p as 100, hence, factor is 8.)

c) 59 * 57 (take b s as 60 and b p as 10, hence, factor is 6.)

ANSWERS: a) 63744, b) 631989, c) 3363

06

Chapter 14: Multiplication Urdhya Tiryak

Okay, this is nothing

but CROSS-MULTIPLICATION.

Suppose you have to multiply:

ab * xy

Then, ab * xy

= ax/ (ay + bx)/ by

Check the short sketch out:

a b

x y

= a x / (a y + b x) / b y

B l u e A r r o w i s t h e f i r s t s t e p . (l e f t - m o s t d i g i t) B l a c k A r r o w i s t h e s e c o n d s t e p . (M i d d l e d i g i t s) R e d A r r o w i s t h e l a s t s t e p . (r i g h t - m o s t d i g i t s)

For eg: 89 * 43

= 32/(36 + 24)/27

= 32/60/27

= 3827

TRYTHESE :

a) 4 3 * 9 8

b) 3 4 * 1 9

c) 8 9 * 7 1

ANSWERS : a) 4 2 1 4 , b) 6 4 6 , c) 6 3
1 9 . 1 6

For three digit this goes like:

abc * xyz

= ax/(ay + bx)/(az + by + cx)/(cy + bz)/ bz

a b c

x y z

= ax/(ay + bx)/(az + by + cx)/(cy + bz)/ bz

Blue Arrow is the first step. (left-most digits). Green Arrow is the second step. (After-left-most digits). Black Arrow is the third step. (Middle-most digits). Orange Arrow is the fourth step. (After-middle-digits). Red Arrow is the last step. (right-most digits).

For eg: 539 * 934

= 45/(15 + 27)/(81 + 9 + 20)/(12 + 27)/36 = 45/42/110/39/36

= 503426

TRYTHESE: a)234*543 b)789*392 c)930*101

ANSWERS: a)127062,

b)309288, c)93930.17

07

DIVISION

So, in the short division, which is shown in this book, the division is done with flags (superscript numbers of the divisor). Let's do the identity: Let abcde be a three-digit number and let it be the dividend and let the divisor be xy. And such is the case that abc is divisible by xy then,

If abc / xy= FG

Note:

a) p/b means a number whose first digit is p and second digit is b.

b) The flag are the digits of the divisor, excluding the first digit.

xy a p b : c

- y F - y G

(p / b - y F) 0

- F x - G x

F G : 0

Step 1: divide the first digit by the nearest multiple of x (first digit of the divisor). Step 2: For some remainder p we write it on the side of the second digit of the dividend (b) and then we first multiple the flag (the last remaining digits of the divisor) and subtract from p/b

Step 3: If the dividend is divisible by the divisor, then we continue the multiple the sub-quotients and subtract by the remainder-of-the-previous-digit/nextdigit, until we get 0.

Note : For more than one flag, you have to cross-multiply with the sub-quotient instead of directly multiplying and subtracting.

08

SQUARES

THE CONCEPT OF DUPLEX

Here, the concept of DUPLEX comes. Duplex of a number (xy) is written as D(xy).

Duplex of one-digit number:

D(a) = a^2.

Duplex of two-digit number:

D(ab) = 2ab.

Duplex of three-digit number: D(abc) = 2ac + b^2.

Duplex of four digit number:

D(abcd) = 2(ad + bc).

Duplex of five-digit number:

D(abcde) = 2(ae + bd) + c^2.

Pre-requisites are done.

FORMULA for SQUARES

Now, we are gonna speedily

do the squares of any number.

Square of one-digit number, suppose a

= D(a) = a^2

Square of two-digit number, suppose ab

= D(a)/D(ab)/D(b)

Square of three-digit number, suppose abc

= D(a)/D(ab)/D(abc)/D(bc)/D(c)

Square of four-digit number, suppose abcd

=D(a)/D(ab)/D(abc)/D(abcd)/

D(bcd)/D(cd)/D(d)

Square of five-digit number, suppose abcde

128

=D(a)/D(ab)/D(abc)/D(abcd)/D(abcde)/
D(bcde)/D(cde)/D(de)/D(e)

EXAMPLES

CALCULATE :

a) (49)^2

= D(4)/D(49)/D(9)

= 16/72/81

= 2401

b) (897)^2

=   D(8)/D(89)/D(897)/D(97)/D(7)   =
64/144/193/126/49

= 804609

c) (23894)^2

=   D(2)/D(23)/D(238)/D(2389)/D(23894)
/D(3894)/D(894)/D(94)/D(4)

= 4/12/41/84/134/168/145/72/16

= 570923236

TRY THESE :

a) (1 3) ^ 2

b) (4 9 2 3) ^ 2

c) (8 4 9 2 2) ^ 2

ANSWERS : a) 1 6 9 , b) 2 4 2 3 5 9 2 9 ,
c) 7 2 1 1 7 4 6 0 8 4 .2 4

09

SQUARE ROOT

Chapter 15: Theory

In this process, there are sub-divisions. So, the first step depends on the number of digits your number has. If it is odd, you consider the first digit. If it's even you take the first two digits into consideration. Suppose you can to find the root of a fivedigit number abcde. Then,

a β / b γ / c θ / d G / e

- D (x) - D (x y) - D (x y 0)

F G 0

2 α - α ^ 2 - 2 α x - 2 α y

α x y 0 0

N O T E :

a) a , β / b , γ / c , e t c . a r e a l l t e r m e d a s s u b - d i v i d e n d s .

b) α (a l p h a) , x a n d y a r e a l l t e r m e d a s s u b - q u o t i e n t s .

c) D (x) , D (x y) , e t c . a r e t h e D u p l e x o f t h e r e s p e c t i v e n u m b e r s . F o r D u p l e x , s e e t h e S q u a r e s p a r t .

d)F = $(\gamma/c - D(x))$; and G = $(\theta/d - D(xy))$

e)Generally, if a number is a perfect square of some number, then, the last digits are of the sub-divisions add up to 0.

f) Sometimes you have to adjust the subquotients such that (remainder/nextdigit − D(previous quotient)) > -1

THE STEPS

These are all variables and I will tell you what happens step-by-step.

Step 1: We take into consideration of the first digit because abcde is odd number. Then we find the highest perfect square less than a which is α here. The remainder of a − α is β. And the number becomes the second digit of the gross

dividend (temporary sub-dividend).

Step 2 : We now write 2α on the

right, and we make it our main divisor.

So, again for some number x, 2αx

becomes the highest number lesser than

β/b which when subtracted we get γ as

the remainder and γ is made the second

digit of the third number c (the third

sub-dividend is now γ/c). And down

that, we write the quotient x for the

second sub-division.2 8 Step 3: Now we take in the Duplex of the second quotient which is x, and subtract with γ/c. we get F (see notes) and then we do the same thing, find out the multiple of the main divisor and write the number for which the multiple is the closest in the next-quotient and subtract the multiple with y/c. and repeat the same steps until we have 0 as the remainder between remainder/nextdigit-of-the-dividend and D(previous quotient).

Generally, the last digit is gonna be zero if it is actually divisible.

For example : Find the root of 1 225. As it is an even digit number, we take the first two digits in to consideration.

1232256-9-30-25350

So, the root of 1225 is 35.

TRY THESE :

Find the root of : a) 2025

b) 14641

c) 1373584

ANSWERS: a) 45, b) 121, c) 1172
.

CUBES

Okay generally this is very easy. Suppose you want to find the cube of a number ab. Then,

$(ab)^3 = a^3 a^2.b a.b^2 b^3 + 2 a^2.b + 2a.b^2$

It's that

SIMPLE.

NOTE: You have to memorize perfectly the cubes of the one-digit numbers.

For example: $(58)^3$

$= 125 200 320 512$

$+ 400 + 640$

$= 125/600/960/512 = 195112$

Okay that a/b/c/d = xyz is due to adding and also CARRYING NUMBERS. Don't worry it's gonna be okay.

TRY THESE :

Find the cube of:

a) 51

b) 4.6

c) 32

ANSWERS: a) 132651, b) 97.336, c) 32768.

Chapter 16: Cube Roots

Okay this is the fastest, if you have it inside of your head the cubes of onedigit numbers. Generally, we do this for 6-digit numbers but everything's fine. Let abcdef be a six-digit number, then divide the number into two groups, three digits each separated by a colon :

abcdef = abc : def = x:y

y is found out by seeing the last digit f, because none of the last digits of the cubes of the one-digits are the same. And x is actually the nearest one-digit number's cube which is lesser than abc. For example:

175616 = 175: 616 = 5:6 = 563

See the cube of 6 is 216 which ends with 6 and the cube of 5 is 125, which is the closest to 175.

Hence the answer.⏹

TRYTHESE :

Find the cube root of : a) 238328

b) 157464

c) 704969

ANSWERS : a) 62, b) 54, c) 89.

TRIGONOMETRY

(values of half and double angles)

Okay, this is nothing but getting out the values of half-angles and double angles if you are given the main angles' value. This is done through the triplets ($c^2 = a^2 + b^2$).

The triplet for the angle A is written as: a (base), b (perpendicular),

c (hypotenuse).

Therefore, $\sin A = b / c$. And all the other ratios likewise.

Then the triplet for the angle 2A is: $(a^2 - b^2)$, $2ab$, c^2

And the triplet for angle A/2 is:

$(a + c)$, b, $\{(a + c)^2 + b^2\}^{(1/2)}$

For example:

value of cos A = 12/13 and we have to find the value of sin 2A, then:

a, b, c = 12, b, 13.

We can easily find that b = 5

Therefore, a, b, c = 12, 5, 13

Then converting the triplets for angle 2A: 119, 120, 169.

Now as

sine   =   perpendicular/hypotenuse. Therefore, sin 2A = 120/169.

T R Y T H E S E :

a) $Tan A = 16/63$, then find $sin 2A$. b) $Tan A = 5/12$, then find $cos 2A$.

ANSWERS: a) $2016/4225$, b) $119/169$.

13

MISHRANK

THE CONCEPT

This is the last chapter of this word document. Well, this concept is of changing a number like 68 into 72, for convenient operations. This is called Mishrank. Mishrank has a negative value. And I'll explain this with an example. If the last digit is 6,7, 8 & 9 then to convert it into its Mishrank, subtract the last digit from 10 and add 1 to the first digit. To convert back, to the same thing but now, subtract 1 from the first digit. We show the Mishrank with a bar overhead. Example: 89 = 91, and 154 = 146.

But the main thing while doing

operations with Mishrank, is keeping in mind which are Mishrank. Because in the end you have to subtract and convert in back into original number.

EXAMPLES 1

S u p p o s e 6 9 * 4 8

6 9 = 7 1 .

4 8 = 5 2 .

T h e r e f o r e , 6 9 * 4 8 = 7 1 * 5 2 (T h i s i s a l o t e a s i e r t o m u l t i p l y , r i g h t ? ▨) 7 1

5 2

= 3 5 : 1 9 : 2

= 3 4 : 9 : 2

= 3 3 : 1 : 2 = 3 3 1 2 .

See. As Mishrank has a negativ e value therefore, we have to su btract to the next digit.

EXAMPLE 2

Mishrank can also be used in various other operations. This is very useful concept which can be used to speed up all your calculations. Example: $(89)^3$

89 = 9 1 .

Hence, $(89)^3$

= $(91)^3$

= 7 2 9 8 1 9 1

16218

= 7 2 9 : 2 4 3 : 2 7 : 1 = 7 0 5 : 1 : 7 : 1 = 7 0 4 9 6 9

TRY THESE : a) 779^2

b) 9 7 * 3 8

A N S W E R S : a) 6 0 6 8 4 1 , b) 3 6 8 6 .

14

AUXILLIARY FRACTIONS

This method is used for divisors which ends with 9 like 19, 39, 49, etc. For

example: 5/29

Here, first we add 1 to the divisor and then remove the zero from the denominator by placing a decimal point in the numerator. The final form becomes 0.5/3

Now, we would do the step-by-step division.

Divide 5 by 3, (q) = 1 and (r) = 2. Write like this:

0.5/3 = 0. 1

2

Now, we need 21 by 3 (diagonal number formation with quotient and remainder) and we continue to get the required level of accuracy.

0.5/3 = 0. 1 7 2 4 1 3 7 9

2 0 1 0 1 2 2 0 So, 5/29 = 5/30 = 0.5/3 = 0.17241379

First, we subtract 1 from both the numerator and denominator; and modify by removing the zero from the

denominator. For example: 4/21 = 3/20 = 0.3/2.

Divide 3 by 2, (q) = 1 and (r) = 1; The difference of quotient from 9 is 8. Write like this:

8

0.3/2 = 0. 1

1

Now divide 18 instead of 11 (what we did in the previous case). And we continue to do it to get the required level of accuracy.

8 0 9 6 2 3 8 0 0.3/2 = 0. 1 9 0 4 7 6 1 9 1 0 0 1 1 0 1 0

Therefore, 4/21 = 0.19047619

Here, add 2 to the denominator and then remove the zero and place a decimal point in the numerator at the

appropriate position.

Since the last digit of the divisor is 8 which has a difference of 1 from 9, we multiply the quotient by 1 and add to the base dividend at each stage to compute the gross dividend.

Let's take the case of 5/28.

5/28 = 0.5/3

0.5/3 = 0.1 7 8 5 6

2 1 0 1 2

 In the second division, we take 7 from the quotient and add it to the formed 17 to get 24. And we continue like this. 5/28 = 0.17856

Here, add 3 to the denominator and then remove the zero and place a decimal point in the numerator at the appropriate position.

Since the last digit of the divisor is 7 which has a difference of 2 from 9, we multiply the quotient by 2 and add to the base dividend at each stage to compute the gross dividend.

 Let's take the case of 5/27.

 5/27 = 0.5/3

 0.5/3 = 0.1 7 13 19 25 31

2 2 2 2 2 2

In the second division, we take 7 from the quotient and multiply it by 2 to get 14 and add it to the formed 27 to get 41. And we continue like this.

And at the end, we add up all the things. Hence, 5/27 = 0.185181

Do the same thing, just multiply the quotient by 3 this time.

6 / 7 6 = 0 . 6 / 8

0 . 6 / 8 = 0 . 0 7 8 9 4 7 3 8 6 4 4 0 4 0 4 4

Therefore, 0.6/8 = 0.07894738

For other divisors like 2,3,4 and 5, multiply them with some number to get your convenient divisor. Multiply 2 by 5 or 4. Multiply 2 by 3 or 7. Multiply 4 by 5 or 7. Multiply 5 by 2. Multiply 6 by 5 or 3. That's all.

15

OSCULATORS

Chapter 17: POSITIVE OSCULATORS

An osculator is a number define
d for any number ending in 9 or 1
.

Generally used to check divisi
bility by numbers 21, 79, etc.

Positive osculator is 9 and neg
ative osculator is 1.

The osculator for 19 is 2,

for 29 it is 3, for 79 it is 8, etc.

You first drop the 9 in case of

positive osculator and then st
art from the right-hand digit of
the dividend. Multiply the digi
t by the osculator and then add
next digit of the dividend. And
multiply it again by the osculat
or. Subtract the

maximum possible multiples of the divisor from it. And continue if there are more digits. IF the end-product is 0 or the digit itself then, it is a multiple of the divisor. And the divisibility is True.

EXAMPLES

Example: Check the divisibility of 437 by 19.

Osculator of 19 is 2. Then first we start from the rightmost digit which is 7. 7*osculator=14; 14+next digit=17; As 17<19, hence we skip the subtraction step here. Then, 17*osculator=34; 34+next digit=38. Now 38−2(19)=0. Hence, 437 is divisible by 19.

NEGATIVE OSCULATORS

Generally, they are the ones which end with 1. Eg: 21, 41, 91, etc.

Osculator of 21 is 2, of 51 is 5, etc. First, we drop 1 and then mark the alternate digits with Mishrank. We start from the rightmost digit and then do the same thing. Multiply with osculator and add the next digit. But here, as it is Mishrank, therefore it would work as

subtraction because Mishrank has a negative value.

For example: 3 + 2 = 1;

So, we do the same thing like this and if it is 0 or a multiple of the divisor itself, the divisibility is true.

EXAMPLE

Example: Check the divisibility of 2793 by 21.

The osculator is 2 for 21.

Now we mark the alternate digits by Mishrank.

2793

Now, 3 * osculator = 6; 6 + 9 = 3; 3 * osculator = 6; 6 + 7 = 1; 1 * osculator = 2; 2 + 2 = 0.

Therefore, 2793 is divisible by 21.

Note: it can also be done if you take the other set of dividends as Mishrank. What I'm saying is that it would be the same if we took

2793 instead of 2793

Numbers like 3 and 7 can also be multiplied (3 by 7 and 7 by 3) to use as osculators. That's all.

16

IN THE END

So, that's it. Thanks for seeing and using this▯. I'm glad that I can be of your help. It's awesome to give out so much knowledge. It was a really good book. And all of it is in is there, except two parts (Simultaneous equations and Applications of Vedic Maths)▯. That Application ... is nothing but exercises and word problems with all those tricks we have learnt and the Simultaneous

Equations has nothing much.

The main parts are all here.

Hope you like it.

Try to share it as much as possible so that others can also use this.

You can interact with me on Instagram, Facebook, Twitter and Pinterest.

@thoughts.and.realizations I also run a blog called

"The Teenager's Diary." This is SHADOW SPARKLING signing off.⍰

You can interact with me on Instagram, Facebook, Twitter and Pinterest.

thoughts.and.realizations

I also run a blog called "The Tee
nager's Diary." This is SHADO
WSPARKLING

signing off.